FLORA OF TROPICAL EAST AFRICA

ULMACEAE

R. M. Polhill

Trees or shrubs, sometimes armed. Leaves simple, alternate; blade often unequal-sided. Stipules lateral, usually free, often small and caducous. Flowers small, polygamous, solitary or in axillary cymes or clusters. Sepals 4–8, free or shortly united, imbricate or valvate, persistent. Petals absent. Stamens the same number as the calyx-lobes and opposite them or a few more, inserted at the base of the calyx, erect in bud; anthers 2-thecous, opening lengthwise. Ovary of 2 connate carpels, 1–2-locular; styles 2, divergent; ovule solitary, pendulous from near the top, anatropous. Fruits either compressed, dry and ± winged or appendiculate, or thinly fleshy. Seeds without endosperm; embryo straight or curved.

About 14 genera and 120 species, mostly tropical and north temperate. The family is rather poorly represented in Africa, but the species which do occur are mostly widespread across the continent.

Fruit a flat stipitate samara with broad membranous wings; flowers in the axils of fallen leaves of the previous season's growth; stamens usually a few more in number than the sepals; leaf-blades ± equal-sided, entire, penninerved . . . . 1. **Holoptelea**

Fruit a thinly fleshy drupe; flowers mostly in the axils of the current year's shoots or precocious; stamens equal in number to the sepals:

- Plants unarmed (in the African species); stipules free; styles less than 5 mm. long:
  - Calyx-lobes of the ♂ flowers imbricate; styles 2–5 mm. long; leaf-blades entire, coarsely toothed or serrate in the upper two-thirds . . . . 2. **Celtis**
  - Calyx-lobes of the ♂ flowers induplicate-valvate; styles 0·5–1(–2) mm. long; leaf-blades serrate almost from the base . . . . . 3. **Trema**
- Plants armed with axillary spines; stipules united along one margin; styles 7–20(–30) mm. long . . 4. **Chaetacme**

1. HOLOPTELEA

Planch. in Ann. Sci. Nat., sér. 3, 10: 259 (1848)

Deciduous trees, unarmed. Leaf-blades penninerved, ± equal-sided, punctate below with small rounded cystoliths. Inflorescences congested cymose with ♂ and ☿ flowers. Sepals 4–6(–8), shortly united at the base. Stamens equal in number to the sepals or usually a few more; anthers pubescent. Ovary compressed, 2-locular, but only one ovule matured; styles spreading, unbranched, persistent. Fruit a stipitate samara, with broad rounded wings fused to the lower part of the styles.

Two species, one African, the other Asiatic.

FIG. 1. *HOLOPTELEA GRANDIS*—**1,** mature leaf, × 1; **2,** flowering branchlet, × 1; **3,** ♂ flower, × 8; **4,** ⚥ flower, × 3; **5,** anther, × 16; **6,** fruiting branchlet, × 1. 1, from *Andrews* 1647; 2–5, from *Brenan* 8982; 6, from *Eggeling* 1130 *in F.H.* 1273.

H. grandis (*Hutch.*) *Mildbr.* in N.B.G.B. 8: 53 (1921); Hauman in F.C.B. 1: 40 (1948); F.P.N.A. 1: 41 (1948); I.T.U., ed. 2: 436, fig. 92 (1952); F.P.S. 2: 255 (1952); F.W.T.A., ed. 2, 1: 593 (1958). Type: Togo Republic, *Kersting* A515 (K, lecto.!)

Large tree, 12–40 m. high, with tall sharp buttresses and smooth grey bark. Twigs corky with many raised lenticels, glabrous to puberulous. Leaf-blades chartaceous or thinly coriaceous, ovate to ovate-oblong, 6·5–17 cm. long, 3·5–9·5 cm. wide, shortly apiculate, rounded to cordate at the base, entire or very obscurely sinuate, glabrous or puberulous on the veins beneath, minutely punctate beneath, scabrous when mature, penninerved with 5–8 veins on each side of the midrib; petiole 5–12 mm. long. Stipules linear, 4–5 mm. long, pubescent, caducous. Inflorescences in the axils of fallen leaves of the previous season's growth, congested cymose with many clustered ♂ flowers and fewer ⚥ flowers above on slender 4–8 mm. long pedicels. Calyx-tube very short; lobes 4–6, ± 1 mm. long. Stamens 7–12. Samara shortly stipitate, suborbicular to obovate, 3·5–4·5 cm. long, 3–3·5 cm. wide, with papery and radially veined wings, subglabrous; styles 4–6 mm. long; nutlet in the lower half of the wings, 9–13 mm. long, 6–8 mm. wide. Fig. 1.

UGANDA. Bunyoro District: Budongo Forest, Feb. 1933, *Eggeling* 1139 *in F.H.* 1273!; Mbale District: Tororo, Nov. 1941, *Dale* U202!; Mengo District: Mabira Forest, Oct. 1904, *Dawe* 172!

DISTR. U1–4; Sudan and Congo Republics to West Africa

HAB. Lowland rain-forest, also drier deciduous and riverine forest; 750–1200 m.

SYN. *Hymenocardia grandis* Hutch. in K.B. 1911: 184 (1911) & in F.T.A. 6(1): 649 (1912); Mildbr. in Z.A.E.: 449 (1912)
[*Holoptelea integrifolia* sensu Rendle in F.T.A. 6(2): 2 (1916), *non* (Roxb.) Planch.]

NOTE. *H. grandis* is closely related to the Asiatic species, *H. integrifolia* (Roxb.) Planch., but the nutlet is situated nearer the base of the wings, the inflorescences are usually fewer-flowered and the plant is altogether less hairy.

Fruiting specimens are superficially similar to certain species of *Hymenocardia*, notably *H. ulmoïdes* Oliv., but may be distinguished by the larger fruits with only one fertile locule, the more persistent styles and the absence of scales.

2. CELTIS

L., Sp. Pl.: 1043 (1753) & Gen. Pl., ed. 5: 467 (1754)

Trees or shrubs, monoecious or rarely dioecious, ± armed. Leaf-blades penninerved or 3–5-nerved from the base, which is often unequal-sided, and with cystoliths, giving a scabrous texture to the mature leaves, often present. Stipules usually small and caducous. Inflorescences cymose or fasciculate, the lower ones usually with many predominantly ♂ flowers, the upper ones with few longer pedicelled ♀, ⚥ flowers. Sepals (4–)5, free or almost so, imbricate. Stamens as many as the sepals; anthers ovate. Ovary sessile, 1-locular; styles ± 1–2-branched. Drupes thinly fleshy, subglobose to ellipsoid; endocarp hard, sometimes ribbed.

Widespread in the tropics and temperate regions mostly of the northern hemisphere. The African species are unarmed.

C. australis L., native to Eurasia, has been cultivated in Tanganyika at Amani—Amani Plant 18, *Greenway* 1589!; Kiumba Plant 4, *Greenway* 2816!

NOTE. Two species described from East Africa as *Celtis* are referable to other genera: *C. ilicifolia* Engl. is *Populus ilicifolia* (Engl.) Rouleau and *C. polyclada* Peter is a synonym of *Ziziphus pubescens* Oliv.

Styles unbranched; ripe fruits* usually less than 6 mm. long:
- Leaf-blades serrate in the upper two-thirds, usually pubescent at least on the nerves beneath, with the basal lateral nerves ascending well into the upper half and the upper prominent lateral nerves 1–2 on each side of the midrib; fruits subglobose, pubescent, on 10–25 mm. long pedicels . . . . 1. *C. africana*
- Leaf-blades entire, soon glabrescent, with the basal lateral nerves extending little into the upper half and the upper prominent lateral nerves 3–6 on each side of the midrib; fruits conic-ovoid, glabrous, on 4–8 mm. long pedicels . 2. *C. durandii*

Styles bifid or conspicuously 1–2-branched; ripe fruits more than 6 mm. long:
- Leaf-blades 3-nerved from the base, with the basal lateral nerves extending little into the upper half and the upper prominent lateral nerves 3–6 on each side of the midrib:
  - Ovary subglabrous; fruits pointed ellipsoid, 4-angled or -ribbed when dry; leaf-blades elliptic or more commonly obovate to obovate-oblong, usually coarsely toothed towards the apex, soon glabrous except for a few hairs on the nerves beneath . . 4. *C. mildbraedii*
  - Ovary densely hairy; fruits ovoid, not angled or ribbed; leaf-blades elliptic to oblong-ovate, entire or toothed, rather persistently pubescent beneath, at least on the nerves . . 5. *C. zenkeri*
- Leaf-blades 3–5-nerved from the base, with the main basal lateral nerves extending into or almost into the acuminate tip and the upper prominent lateral nerves 1–2(–3) on each side of the midrib:
  - Leaf-blades ovate, broadly rounded to subcordate at the base, 4·5 × 3·5 – 8 × 5 cm., hairy in the angles of the nerves beneath, scabrous. 3. *C. integrifolia*
  - Leaf-blades elliptic to elliptic-oblong, 8·5 × 4 – 17·5 × 8 cm., glabrous or subglabrous, not scabrous:
    - Ovary glabrous; fruits up to 12 × 10 mm.; leaf-blades entire or coarsely toothed in the upper half . . . . . 6. *C. wightii*
    - Ovary pubescent; fruits 15 × 13 – 20 × 18 mm.; leaf-blades entire . . . 7. *C. adolfi-fridericii*

1. **C. africana** *Burm. f.*, Prodr. Fl. Cap.: 31 (1768); Brenan in Mem. N.Y. Bot. Gard. 9: 75 (1954); Verdoorn in Fl. Pl. Afr. 31, t. 1210 (1956); F.W.T.A., ed. 2, 1: 592 (1958); K.T.S.: 574, fig. 104 (1961); F.F.N.R.: 22 (1962); Polhill in K.B. 19: 139 (1964). Type: t. 88 in Burm., Rar. Pl. Afr. (1739)

Deciduous tree, 5–35 m. tall, with smooth grey bark and often slight horizontal annular ridges (ring marked). Young twigs tawny pubescent to tomentose. Leaf-blades ovate to ovate-lanceolate, 5–10·2 cm. long, 2–5·5 cm. wide, acuminate, rounded and a little unequal-sided at the base,

* Dimensions throughout are of dried fruits; in examining fresh drupes the hard endocarp alone should be measured.

serrate in the upper two-thirds, pubescent, ultimately subglabrous except on the nerves beneath, ± scabrous, with the basal lateral nerves extending well into the upper half and the upper prominent lateral nerves 1–2 on each side of the midrib; petiole 1–5 mm. long. Stipules linear, 4–6 mm. long, pubescent. Cymes in the lower leaf axils and at the nodes below of 3–15 clustered ♂ flowers with 1·5–4 mm. long pedicels; uppermost cymules of 1–several ⚥ flowers with long slender 10–17 mm. long pedicels; intermediate ones polygamous. Sepals 4–5, 1·5–3 mm. long, pubescent. Ovary densely hairy; styles unbranched, 2·5–4 mm. long. Fruits on 10–25 mm. long pedicels, subglobose, ± 5 mm. long, orange, pubescent.

UGANDA. Acholi District: SE. Imatong Mts., Aringa R., 7 Apr. 1945, *Greenway & Hummel* 7307!; near Toro, on Hoima road, 1915, *Fyffe* 11!; Mengo District: Mabira Forest, Mulange, Aug.–Sept. 1919, *Dummer* 4301!

KENYA. Nakuru District: Njoro, *Battiscombe* 508!; N. Kavirondo District: Kakamega Forest Station, 17 Sept. 1949, *Maas Geesteranus* 6267!

TANGANYIKA. Mwanza District: Rubondo I., 27 Oct. 1956, *Gane* 81!; Arusha, track to Usa Sawmills, 22 Nov. 1957, *Hughes* 124!; Mpwapwa District: Kiboriani Mts., 3 Oct. 1938, *Greenway* 5795!

ZANZIBAR. Zanzibar I., Kombeni cave-well, 7 Sept. 1930, *Vaughan* 1491!

DISTR. **U**1–4; **K**3–5; **T**1, 2, 5, 7; **Z**; widespread from Arabia to the Cape Province of South Africa and from the Sudan Republic to Nigeria and Angola

HAB. Dry evergreen and riverine forest, also upland rain-forest; 30–2400 m.

SYN. *C. kraussiana* Bernh. in Flora 28: 87 (1845); V.E. 3(1): 12 (1915); Rendle in F.T.A. 6(2): 3 (1916); F.D.O.-A. 2: 64 (1932); T.S.K.: 84 (1936); Hauman in F.C.B. 1: 43 (1948); F.P.N.A. 1: 43 (1948); T.T.C.L.: 624 (1949); I.T.U., ed. 2: 434, fig. 89b (1952); F.P.S. 2: 253 (1952); E.P.A.: 5 (1953). Type: South Africa, Cape Province, *Krauss* (G, iso.!)

C. holtzii Engl., V.E. 3(1): 12, fig. 6E (1915). Type: Tanganyika, Mwanza, *Holtz* 1591 (B, holo.!)

C. kraussiana Bernh. var. *stolzii* Peter, F.D.O.-A. 2: 64 (1932); T.T.C.L.: 624 (1949). Type: Tanganyika, Rungwe District, *Stolz* 1708 (B, holo.!, BM, K, iso.!)

NOTE. A well defined, but rather variable species, closely related to several from Eurasia, including *C. australis* L.

2. **C. durandii** *Engl.* in N.B.G.B. 3: 22 (1900) & in Z.A.E.: 179 (1911) & V.E. 3(1): 12, fig. 6D (1915); Rendle in F.T.A. 6(2): 4 (1916); F.D.O.-A. 2: 65 (1932); Hauman in F.C.B. 1: 42 (1948); F.P.N.A. 1: 43 (1948); T.T.C.L.: 624 (1949); F.W.T.A., ed. 2, 1: 592 (1958); F.F.N.R.: 431 (1962); Polhill in K.B. 19: 140 (1964). Types: Tanganyika, Iringa/Kilosa District, Usagara, *von Trotha* 171 (B, syn.!) & Congo Republic, Bas-Congo, *Dupuis* (B, syn.!, BR, isosyn.!)

Much branched monoecious or usually dioecious deciduous tree, 5–25 m. tall, with smooth light grey bark; dead wood with a peculiar foetid smell. Twigs pubescent or rarely subglabrous; lenticels prominent on the older branchlets. Leaf-blades membranous or chartaceous, oblanceolate, oblong-elliptic or ovate, 5·5–16·5 cm. long, 2·2–7·5 cm. wide, long-acuminate, cuneate to rounded at the base, slightly unequal-sided, entire or rarely with a few coarse teeth, glabrescent, 3-nerved from the base, with the basal lateral nerves extending little into the upper half and the upper prominent lateral nerves 3–6 on each side of the midrib; petiole 5–8 mm. long. Stipules linear-oblong, 4–5 mm. long, pubescent, caducous. Cymules of ♂ flowers often precocious, crowded, with short 2–6 mm. long pedicels; ♀ and ⚥ flowers axillary or at the nodes below, few, with 3–7 mm. long pedicels. Sepals 4–5, 1·5–2·5 mm. long, pubescent. Ovary glabrous or puberulous; styles unbranched, 2–3 mm. long. Fruits conic-ovoid, 4-angled when dry, 4–6 mm. long, 3–4 mm. across, glabrous. Fig. 2, p. 6.

UGANDA. Bunyoro District: Budongo Forest, Feb. 1935, *Eggeling* 1619 *in F.H.* 1514! & Mar. 1936, *Eggeling* 3068!; Mengo District: Mabira Forest, near Najembe, 14 Apr. 1950, *Dawkins* 570!

FIG. 2. *CELTIS DURANDII*—**1,** flowering branchlet, $\times \frac{2}{3}$; **2,** ♂ flower with one sepal and stamen removed, $\times 8$; **3,** same with aborted ovary more developed, $\times 8$; **4,** sepal and stamen, $\times 12$; **5,** ♀ flower, $\times 8$; **6,** longitudinal section of ♀ flower, $\times 6$; **7,** fruiting branchlet, $\times \frac{2}{3}$. 1–4, from *Kakoire* 79; 5, 6, from *Koritschoner* 1574; 7, from *Wallace* 1201.

KENYA. N. Kavirondo District: Kakamega, 17 Sept. 1949, *Maas Geesteranus* 6266! & June 1933, *Dale in F.D.* 3082!
TANGANYIKA. Kilimanjaro, Lyamungu, 24 Aug. 1932, *Greenway* 3126!; Lushoto District: Makuyuni, 23 Mar. 1936, *Koritschoner* 1574!; Morogoro District: Nguru Mts., Nov. 1954, *Semsei* 1920!
DISTR. **U**2–4; **K**1, 4, 5; **T**2, 3, 6, ? 7; Mozambique, Zambia and Rhodesia to the Cape Province of South Africa, also Congo Republic and Angola to Nigeria and S. Tomé
HAB. Lowland and upland rain-forest; 300–2000 m.

SYN. *C. ugandensis* Rendle in J.B. 44: 341 (1906). Type: Uganda, Entebbe, *Bagshawe* 669 (BM, holo.!)
C. durandii Engl. var. *ugandensis* (Rendle) Rendle in F.T.A. 6(2): 5 (1916); T.S.K.: 84 (1936); Hauman in F.C.B. 1: 43 (1948); F.P.N.A. 1: 43 (1948); I.T.U., ed. 2: 432 (1952); K.T.S.: 574 (1961)

NOTE. A very characteristic and distinct species, most closely related to *C. gomphophylla* Bak., an endemic of Madagascar and the Comoro Is.

3. **C. integrifolia** *Lam.*, Encycl. 4: 140 (1797); Oliv. in Trans. Linn. Soc. 29: 148 (1875); V.E. 3(1): 14, fig. 6A (1915); Rendle in F.T.A. 6(2): 7 (1916); I.T.U., ed. 2: 432 (1952); F.P.S. 2: 251, fig. 87 (1952); F.W.T.A., ed. 2, 1: 592, fig. 171 (1958); Polhill in K.B. 19: 140 (1964). Type: Senegal, *Adanson* 229A (P, holo., K, photo.!)

Large spreading deciduous monoecious tree, up to 25 m. high, with tall narrow buttresses; bark light grey, smooth or scaly. Young twigs pubescent. Leaf-blades broadly ovate, 4·6–7·8 cm. long, 3·5–4·8 cm. wide, shortly acuminate, rounded to subcordate at the base, unequal-sided, usually entire, sometimes toothed in the upper half, scabrous, subglabrous with a few hairs on the veins beneath and in the vein-angles, 3–5-nerved from the base, with the basal lateral nerves ascending well into the upper half and the upper prominent lateral nerves 1–3 on each side of the midrib; petiole 4–6 mm. long. Stipules linear, 4–5 mm. long, pubescent. Cymes 1–2 cm. long in flower, up to 4 cm. in fruit, axillary or at the nodes below, with many clustered ♂ flowers and, the upper ones in particular, with several terminal ♀, ⚥ flowers on ± 3 mm. long pedicels. Sepals 5, 2–2·5 mm. long, pubescent. Ovary shortly pubescent with a ring of longer hairs at the base; styles 2–3-branched, 3–5 mm. long. Fruits subglobose, 8–11 mm. long, 7·5–10 mm. across, brownish, puberulous or scabrous.

UGANDA. W. Nile District: Laropi Camp, Dec. 1931, *Brasnett* 332!; Acholi District: Madi, 1863, *Grant* 753! & SE. Imatong Mts., Agora, 9 Apr. 1945, *Greenway & Hummel* 7327!
DISTR. **U**1; Sudan Republic, west to Senegal and east to Arabia
HAB. Dry riverine forest and deciduous woodland; 600–1200 m.

4. **C. mildbraedii** *Engl.* in E.J. 43: 309 (1909) & in Z.A.E.: 180, fig. 16E (1911) & V.E. 3(1): 14 (1915); Hauman in F.C.B. 1: 45 (1948); F.W.T.A., ed. 2, 1: 592 (1958); K.T.S.: 576 (1961); Polhill in K.B. 19: 140 (1964). Type: Congo Republic, Orientale, *Mildbraed* 2897 (B, lecto., K, photo.!)

Evergreen or deciduous monoecious tree, 3–45 m. tall, with sharp buttresses and pale smooth bark, scaling in small discs. Young twigs tawny pubescent. Leaf-blades chartaceous to thinly coriaceous, elliptic or more usually obovate to obovate-oblong, 6·5–15 cm. long, 4·5–6 cm. wide, long-acuminate, ± cuneate and unequal-sided at the base, obscurely to distinctly and coarsely toothed in the upper half, glabrescent except for a few hairs on the veins beneath, 3-nerved from the base, with the basal lateral nerves extending little into the upper half and the upper prominent lateral nerves 3–6 on each side of the midrib; petiole 3–7 mm. long. Stipules lanceolate, 4–7 mm. long, pubescent. Cymes short, 5–15 mm. long in flower, up to 30 mm. in fruit, mostly with many clustered ♂ flowers and usually only 1 longer pedicelled ♀ or ⚥ flower at the top, but the uppermost cymules with several

⚥ flowers. Sepals 5, 1–2 mm. long, pubescent. Ovary subglabrous, often with a few hairs in a ring at the base; styles 1–2-branched, 4–5 mm. long. Fruits on 8–13(–20) mm. long pedicels, ovoid-ellipsoid, conspicuously 4-ribbed when dried, 7–10 mm. long, 5–7 mm. across, red, glabrous.

UGANDA. Bunyoro District: Budongo Forest, Nov. 1932, *C. M. Harris* 160 *in F.H.* 1122!; Busoga District: Jinja, Oct. 1931, *C. M. Harris* 35 *in F.H.* 447!; Mengo District: Mabira Forest, near Najembe, 14 Apr. 1950, *Dawkins* 566!

KENYA. N. Kavirondo District: Kakamega, June 1933, *Dale in F.D.* 3083! & Kakamega Forest, June 1961, *Lucas* 145!

TANGANYIKA. Lushoto District: Amani, 16 Nov. 1928, *Greenway* 1008!; Morogoro District: Turiani, Feb. 1939, *Wigg in F.H.* 1411!

DISTR. **U**2–4; **K**5, 7; **T**3, 6; Sudan and Congo Republics to Angola and West Africa, also Rhodesia (Chirinda Forest) and Natal

HAB. Lowland rain-forest; 300–1600 m.

SYN. [*C. soyauxii* sensu Engl. in N.B.G.B. 3: 23 (1900), pro parte; Rendle in F.T.A. 6(2): 5 (1916); T.S.K.: 84 (1936); T.T.C.L.: 624 (1949); I.T.U., ed. 2: 435, fig. 89c, photo. 55 (1952); F.P.S. 2: 251 (1952), *non* Engl. sensu stricto]

C. usambarensis Engl. in E.J. 43: 309 (1909) & V.E. 3(1): 14 (1915); F.D.O.-A. 2: 65 (1932). Type: Tanganyika, E. Usambara Mts., *Zimmermann in Herb. Amani* 853 (B, holo.!)

NOTE. Most closely related to *C. zenkeri*, a species with which it has sometimes been confused.

5. **C. zenkeri** *Engl.* in N.B.G.B. 3: 22 (1900) & V.E. 3(1): 12 (1915); Rendle in F.T.A. 6(2): 6 (1916); Hauman in F.C.B. 1: 45 (1948); F.P.N.A. 1: 44 (1948); T.T.C.L.: 625 (1949); I.T.U., ed. 2: 435, fig. 91a (1952); F.P.S. 2: 251 (1952); F.W.T.A., ed. 2, 1: 592 (1958); Polhill in K.B. 19: 141 (1964). Type: Cameroun, Yaounde, *Zenker & Staudt* 9 (K, iso.!)

Much branched deciduous monoecious tree, up to 10–30 m. tall, buttressed at the base; bark smooth, grey. Twigs tawny pubescent to tomentose. Leaf-blades chartaceous, oblong-elliptic to ovate, 7·5–14 cm. long, 4–7·8 cm. wide, shortly acuminate, broadly cuneate to rounded at the base, unequal-sided, entire or more rarely toothed in the upper half, glabrous above when mature, usually hairy beneath at least on the veins, 3-nerved from the base, with the basal lateral nerves extending a little way into the upper half and the upper prominent lateral nerves 2–4 on each side of the midrib; petiole 5–8 mm. long. Stipules lanceolate, (5–)7–10 mm. long, pubescent, scarious. Cymes 1–2 cm. long in flower, up to 3 cm. in fruit, axillary or at the nodes below; lower ones with many clustered ♂ flowers, often with 1–2 ♀ or ⚥ flowers at the top; upper ones with several longer pedicelled ⚥ flowers. Sepals 4–6, 1·5–2·5 mm. long, pubescent. Ovary densely tawny pubescent; styles 1–2-branched, 2·5–4 mm. long. Fruits on 3–5 mm. long pedicels, subglobose or ovoid, 7–8 mm. long, 6–6·5 mm. across, red, pubescent or subglabrous.

UGANDA. Bunyoro District: Budongo Forest, Apr. 1933, *Eggeling* 1190 *in F.H.* 1292! & Dec. 1934, *Eggeling* 1566 *in F.H.* 1480!; Mengo District: Mabira Forest, Oct. 1904, *Dawe* 173!

TANGANYIKA. Mbulu District: Lake Manyara National Park, Mto wa Ukindu, 22 Nov. 1963, *Greenway & Kirrika* 11184!; Tanga District: 11 km. E. of Korogwe, Magunga Estate, 26 June 1953, *Drummond & Hemsley* 3016!; Morogoro District: Turiani, 29 Dec. 1947, *Wigg in F.H.* 2213!

DISTR. **U**2, 4; **T**2, 3, 6; Sudan and Congo Republics, also Angola to West Africa

HAB. Lowland rain-forest; 250–1200 m.

SYN. *C. stuhlmannii* Engl. in N.B.G.B. 3: 23 (1900) & in Z.A.E.: 181 (1911) & V.E. 3(1): 12, fig. 6B, C (1915); F.D.O.-A. 2: 64 (1932). Type: Tanganyika, Uluguru Mts., *Stuhlmann* 8703 (B, lecto.!, K, isolecto.!)

C. soyauxii Engl. in N.B.G.B. 3: 23 (1900), sensu stricto; Hiern, Welw. Cat. Afr. Pl. 4: 1028 (1900), pro parte. Types: République du Congo, Loango, *Soyaux* 202 (K, isosyn.!) & Angola, Cuanza Norte, *Welwitsch* 6285 (BM, K, isosyn.!)

NOTE. A single leaf from the forest floor on the Nyambeni Hills, Meru District, Kenya (*Verdcourt & Polhill* 2977A) appears to be this species and if so extends considerably the known range in East Africa.

6. **C. wightii** *Planch.* in Ann. Sci. Nat., sér. 3, 10: 307 (1848); Polhill in K.B. 19: 141 (1964). Types: India, *Wight* 85 & *Wight* & Ceylon, *Walker* 214 & *Gardner* (all K, syn. !)

Much branched evergreen monoecious tree or shrub, 3–20 m. tall, often with short sharp buttresses; bark smooth, grey. Young twigs puberulous or subglabrous. Leaf-blades chartaceous to coriaceous, elliptic to elliptic-oblong, (5–)8·5–17·5 cm. long, 3·5–7·8 cm. wide, acuminate, broadly cuneate to rounded or subcordate at the base, ± a little unequal-sided, entire or coarsely toothed in the upper half, glabrous, 3-nerved from the base, with the basal lateral nerves extending practically to the apex and the upper prominent lateral nerves 1–2(–3) on each side of the midrib; petiole 4–16 mm. long. Stipules lanceolate, 3–7 mm. long, shortly produced below the point of attachment, ± pubescent. Cymes 5–30 mm. long in flower, up to 40 mm. long in fruit; lower ones with many clustered ♂ flowers and a few ♀ and ⚥ flowers at the top; upper ones with several ⚥ flowers. Sepals 4–5, 1·5–2·5 mm. long, pubescent. Ovary glabrous or subglabrous with a basal ring of long hairs; styles shortly bifurcate, 1–2–3·5 mm. long. Fruits on 3–7 mm. long pedicels, ovoid, shortly beaked, up to 12 mm. long, 10 mm. across, red, glabrous.

UGANDA. Bunyoro District: Budongo Forest, Apr. 1938, *Eggeling* 3587!; Toro District: Bwamba Forest Reserve, Mungilo, 27 July 1960, *Paulo* 609!; Mengo District: Mabira Forest, Kyagwe, Jan. 1908, *Ussher* 57!

KENYA. Kwale District: 24 km. SW. of Kwale, Mwasangombe Forest, 27 Aug. 1953, *Drummond & Hemsley* 4014! & Mrima Hill, 31 Aug. 1959, *Verdcourt* 2405!

TANGANYIKA. Tanga District: near Korogwe, Magunga Estate, 21 Nov. 1952, *Faulkner* 1065! & 28 Nov. 1952, *Faulkner* 1094!; Morogoro District: Kimboza Forest Reserve, July 1952, *Semsei* 820!; Lindi District: near Nachingwea, Kingupira, Sept. 1957, *Nicholson* 126!

ZANZIBAR. Zanzibar I., Kombeni cave-well, 22 Nov. 1930, *Vaughan* 1687!

DISTR. **U**2, 4; **K**7; **T**3, 6, 8; **Z**; Ethiopia to West Africa and south to Angola and Mozambique, also Madagascar, Mascarene Is. and tropical Asia to Australia

HAB. Lowland rain, swamp and riverine forest; 30–1200 m.

SYN. *C. mauritiana* Planch. in Ann. Sci. Nat., sér. 3, 10: 307 (1848); P.O.A. C: 160 (1895); Rendle in F.T.A. 6(2): 9 (1916); F.D.O.-A. 2: 64 (1932); T.T.C.L.: 624 (1949); K.T.S.: 576 (1961). Type: Mauritius, *Commerson* (K, holo. !)

[*C. prantlii* sensu Engl. in N.B.G.B. 3: 23 (1900), pro parte & in Z.A.E.: 179 (1911) & V.E. 3(1): 12 (1915); Rendle in F.T.A. 6(2): 8 (1916) & 356 (1917); F.W.T.A. 1: 423 (1928); Hauman in F.C.B. 1: 43 (1948); F.P.N.A. 1: 44 (1948), *non* Engl. sensu stricto]

C. brownii Rendle in J.B. 53: 298 (1915) & in F.T.A. 6(2): 10 (1916); I.T.U., ed. 2: 430, fig. 91b (1952); F.W.T.A., ed. 2, 1: 592 (1958). Type: Uganda, Mengo District, Mabira Forest, *E. Brown* 462 (BM, lecto. !)

[*C. philippensis* sensu Leroy in Fl. Madagas. 54: 3, fig. 3/1–4 (1952); Berhaut, Fl. Seneg.: 144 (1954), *non* Blanco]

VARIATION. As treated here the species shows considerable variation in texture, size and toothing of the leaves. The Uganda material generally has at least some leaves coarsely toothed and often rather large, chartaceous and markedly acuminate. The coastal and lowland rain-forest specimens of Kenya and Tanganyika, on the other hand, have typically entire, smaller, more coriaceous and less acuminate leaves. The first form is characteristic of Ethiopia and Congo Republic, but both forms occur again in Cameroun, Angola and West Africa. The second form is found in Mozambique and the Mascarene Is. Both forms may be closely matched by specimens from Ceylon and the Andaman Is., including the syntypes.

7. **C. adolfi-fridericii** *Engl.* in E.J. 43: 308 (1909) & in Z.A.E.: 179, t. 16A–D (1911) & V.E. 3(1): 14 (1915); Rendle in F.T.A. 6(2): 9 (1916); Hauman in F.C.B. 1: 44 (1948); F.P.N.A. 1: 46 (1948); I.T.U., ed. 2: 430, fig. 89a (1952); F.P.S. 2: 253 (1952); F.W.T.A., ed. 2, 1: 592 (1958);

Polhill in K.B. 19: 143 (1964). Types: Congo Republic, Kivu Province, near Ruwenzori, *Mildbraed* 2725 & Orientale, *Mildbraed* 2169 & 2242 (all B, syn.!)

Large semi-deciduous monoecious tree, up to 36 m. tall, with buttresses. Twigs pubescent or almost glabrous. Leaf-blades chartaceous or thinly coriaceous, oblong-elliptic, 10·5–16 cm. long, 5–8 cm. wide, shortly acuminate, rounded and unequal-sided at the base, entire, glabrous or with a few hairs on the veins beneath, 3-nerved from the base, with the basal lateral nerves extending almost to the apex and the upper prominent lateral nerves 1–2 on each side of the midrib; petiole 5–18 mm. long. Stipules small, ± 3 mm. long, shortly produced below the point of attachment, pubescent, caducous. Cymes 1–4 cm. long in flower, up to 6 cm. in fruit, axillary or at the nodes below; lower ones with many clustered ♂ flowers; upper ones with several ⚥ and ♀ flowers at the top. Sepals 5, 1–1·5 mm. long, pubescent. Ovary pubescent with a ring of longer hairs at the base; styles branched, 3–4 mm. long. Fruits ovoid to obovoid, 15–18(–20) mm. long, 13–15(–18) mm. across, red, glabrescent.

UGANDA. Bunyoro District: Bugoma Forest, 26 June 1910, *Dawe* 1022!; Toro District: Bwamba, Kitengya, Aug. 1937, *Eggeling* 3364!
DISTR. U2; Sudan and Congo Republics to Ivory Coast
HAB. Lowland rain-forest; 600–900 m.

NOTE. Closely related to *C. wightii* and sterile specimens are not easy to distinguish, but in the area common to both species, *C. wightii* normally has at least some leaves with coarse teeth.

3. TREMA

Lour., Fl. Cochinch.: 562 (1790)

Trees or shrubs, monoecious or dioecious, unarmed. Leaf-blades penninerved, usually serrate, ± scabrous. Stipules paired, lateral, free. Inflorescences usually congested axillary cymes with ♂ and ♀ or ⚥ flowers. Sepals (4–)5, shortly united, with ♂ buds induplicate-valvate and ♀ buds ± imbricate. Stamens equal in number to the sepals. Ovary sessile, 1-locular; styles short, divaricate or inrolled, unbranched, ± persistent. Drupes small, thinly fleshy; endocarp hard.

About 15 species in the tropics and subtropics.

T. orientalis (*L.*) *Bl.*, Mus. Bot. Lugd.-Bat. 2: 62 (1852); Leroy in Fl. Madag. 54: 10 (1952); Polhill in K.B. 19: 143 (1964). Types: Ceylon, *Herb. Hermann* folio 2: 1 & 4: 71 (BM-SL, syn.!)

Shrub, small or medium sized monoecious or rarely dioecious tree, up to 12 m. tall, with smooth grey bark. Twigs pubescent to tomentose. Leaf-blades oblong-lanceolate to attenuate-ovate, (2–)7·5–14 cm. long, (1·2–)2·3–7·2 cm. wide, acuminate, rounded to cordate at the base, evenly and closely serrate from near the base, glabrous to puberulous above, pubescent to tomentose beneath, scabrous, penninerved; petiole 7–12(–15) mm. long. Stipules lanceolate, 4–7 mm. long, pubescent, caducous. Cymes usually congested, ± 1 cm. long in flower, up to 2 cm. in fruit, many-flowered, mostly ♂, with a few ♀(⚥) flowers at the top. Calyx-tube short; lobes 5, 1–2 mm. long, pubescent. Ovary pubescent; styles inrolled or divaricate, 0·5–1 mm. long, usually persistent. Drupes black, thinly fleshy, ovoid to globose, 3–5 mm. long when dried, glabrous. Fig. 3.

UGANDA. W. Nile District: Arua, Dec. 1939, *Hazel* 415!; Ankole District: Igara, May 1939, *Purseglove* 690!; Mengo District: Mabira Forest, near Najembe, 14 Apr. 1950, *Dawkins* 568!

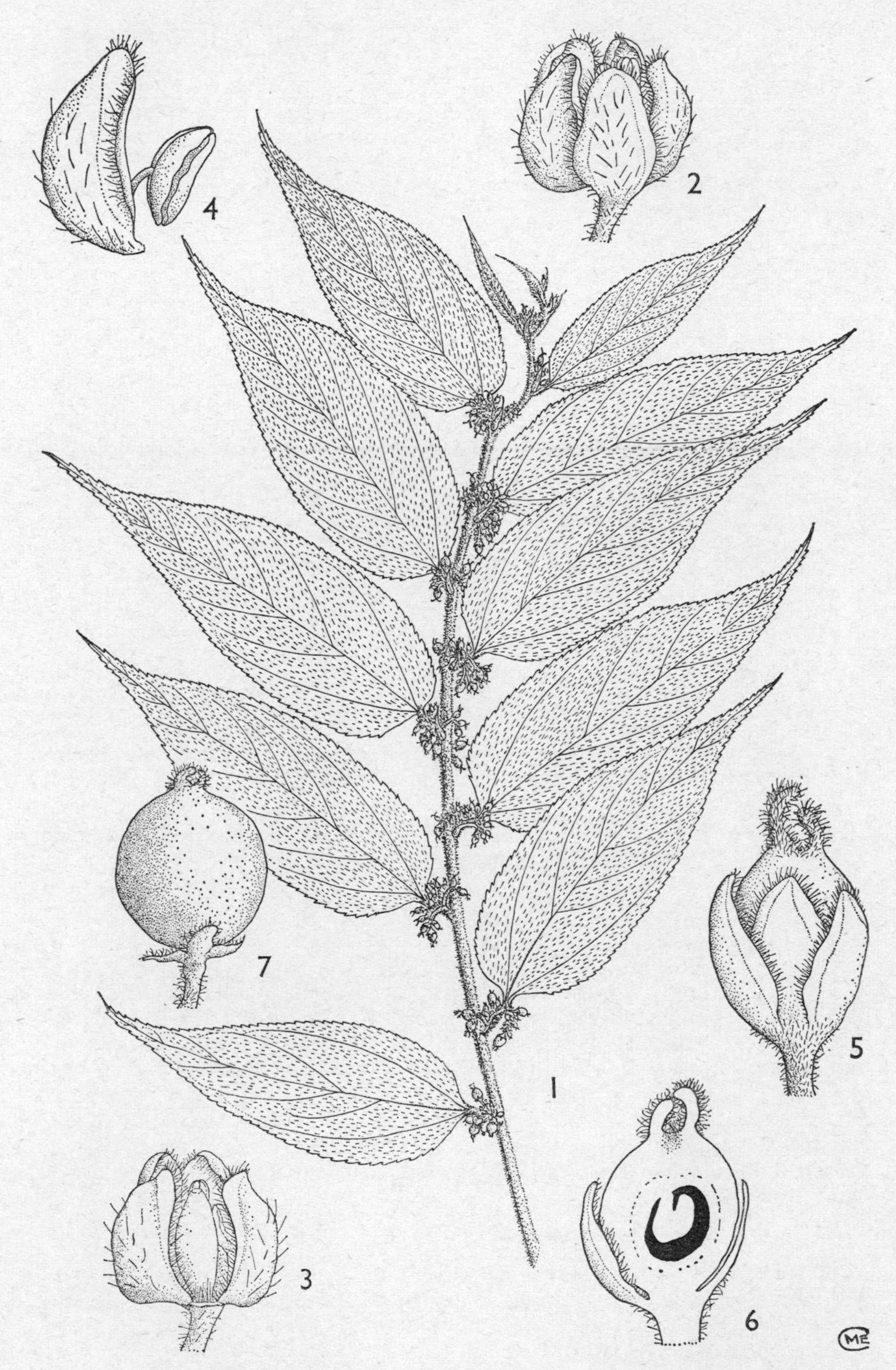

FIG. 3. *TREMA ORIENTALIS*—**1,** flowering branchlet, × ⅔; **2,** ♂ flower, × 16; **3,** ♂ flower with one sepal removed, × 16; **4,** sepal and stamen, × 20; **5,** ♀ flower, × 20; **6,** longitudinal section of ♀ flower, × 20; **7,** fruit, × 6. 1, 5, 6, from *Fundi* 59; 2–4, from *Vaughan* 2636; 7, from *Benedicto* 38.

KENYA. Northern Frontier Province: Mathews Range, Kichich, 23 Dec. 1958, *Newbould* 3561!; Embu, 15 May 1932, *M. D. Graham in A.D.* 1727!; Kilifi District: Sokoke Forest, 27 Feb. 1945, *Jeffery* 103!

TANGANYIKA. Mwanza, 11 July 1933, *C. G. Rogers* 536!; Mpwapwa, 9 Mar. 1930, *Hornby* 198!; Lindi District: Rondo Plateau, 10 Dec. 1955, *Milne-Redhead & Taylor* 7599!

ZANZIBAR. Zanzibar I., Kibwona, 19 Oct. 1930, *Vaughan* 1654!; Pemba I., Chuaka, 10 Oct. 1929, *Burtt Davy* 22441!

DISTR. **U**1–4; **K**1, 3–7; **T**1–8; **Z**; **P**; throughout Africa south of the Sahara, Madagascar, Mascarene Is. and tropical Asia

HAB. Margins of lowland and upland rain-forest, often a pioneer in clearings, also riverine forest; 0–2100 m.

SYN. *Celtis orientalis* L., Sp. Pl.: 1044 (1753)
C. guineënsis Schumach. & Thonn., Beskr. Guin. Pl.: 160 (1827). Type: Ghana, *Thonning* (C, iso.)
Sponia hochstetteri Planch. in DC., Prodr. 17: 198 (1873). Type: Ethiopia, Gojjam, *Schimper* 195 (K, isolecto.!)
Trema guineënsis (Schumach. & Thonn.) Ficalho, Pl. Ut. Afr. Portug.: 261 (1884); P.O.A. C: 160 (1895); V.E. 3(1): 14, fig. 7 (1915); Rendle in F.T.A. 6(2): 11 (1916); F.D.O.-A. 2: 62 (1932); Hauman in F.C.B. 1: 48, t. 8 (1948); F.P.N.A. 1: 46, t. 4 (1948); T.T.C.L.: 625 (1949); U.O.P.Z.: 475, fig. (1949); I.T.U., ed. 2: 438 (1952); F.P.S. 2: 256, fig. 89 (1952); Brenan in Mem. N.Y. Bot. Gard. 9: 76 (1954); F.W.T.A., ed. 2, 1: 592 (1958); K.T.S.: 577 (1961); F.F.N.R.: 24 (1962)
T. guineënsis (Schumach. & Thonn.) Ficalho var. *hochstetteri* (Planch.) Engl., P.O.A. C: 160 (1895) & V.E. 3(1): 14 (1915); Rendle in F.T.A. 6(2): 12 (1916); F.D.O.-A. 2: 63 (1932)

VARIATION. As treated here the species shows considerable variation in the size, shape, texture and indumentum of the leaves and to a lesser extent in the inflorescence-size and style-length, but these appear to be no more than individual variations, which appear throughout its very wide range.

A form from northern Uganda, e.g. W. Madi, Metuli, *Eggeling* 1797!, with a shrubby habit and small leaves, less than 5 cm. long and 1·5 cm. wide with only 2–4 pairs of lateral nerves is more strikingly different. It extends throughout the area of deciduous woodland and bushland from Uganda to West Africa, where collectors have noted that it is very characteristic of this habitat. It has been described as *T. guineënsis* (Schumach. & Thonn.) Ficalho var. *paucinervia* Hauman, but many intermediates do undoubtedly occur in West Africa and it cannot be distinguished on leaf-form from certain South African specimens, including the type of *Sponia glomerata* Hochst. It seems best, therefore, to recognize it as a well marked ecotype without giving it formal taxonomic rank.

4. CHAETACME

Planch. in Ann. Sci. Nat., sér. 3, 10: 266 & 340 (1848)

Trees or shrubs, dioecious or monoecious; branches with axillary spines. Leaf-blades penninerved, minutely punctate on the lower surface, scabrous when mature. Stipules large, united along one margin, enclosing the terminal bud, caducous. Cymes congested, axillary, with many ♂ flowers and rarely an odd ♀ flower at the top; ♀ flowers often solitary, in the upper axils if monoecious. Sepals 5, shortly united at the base, with ♂ buds induplicate-valvate and ♀ imbricate. Stamens equal in number to the sepals. Ovary sessile, 1-locular; styles long, divaricate. Fruits thinly fleshy; endocarp bony, very hard.

One species in Africa and Madagascar.

C. aristata *Planch.* in Ann. Sci. Nat., sér. 3, 10: 341 (1848); P.O.A. C: 160 (1895); V.E. 3(1): 15, fig. 8 (1915); T.T.C.L.: 625 (1949); I.T.U., ed. 2: 436 (1952); F.P.S. 2: 254, fig. 88 (1952); F.W.T.A., ed. 2, 1: 593 (1958); K.T.S.: 576 (1961); F.F.N.R.: 22 (1962); Polhill in K.B. 19: 144 (1964). Type: South Africa, Cape Province, *Drège* (K, isolecto.!)

Shrub or small bushy tree up to 10 m. tall, with drooping, often zigzag branches; bark smooth, grey, becoming fibrous and longitudinally striate.

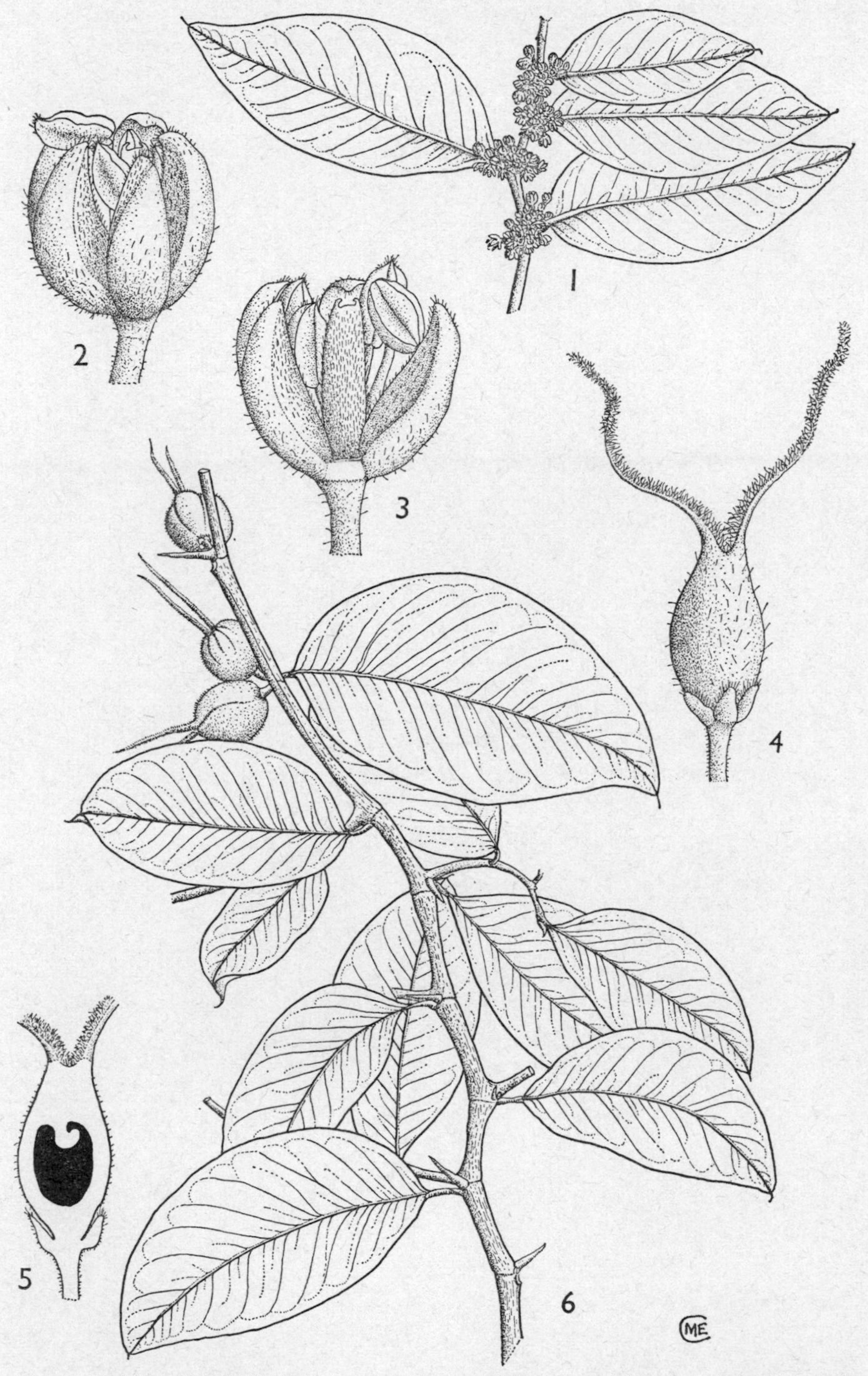

FIG. 4. *CHAETACME ARISTATA*—**1,** ♂ flowering branchlet, × ⅔; **2,** ♂ flower, × 10; **3,** ♂ flower with one sepal removed, × 10; **4,** ♀ flower, × 6; **5,** longitudinal section of ♀ flower, × 10; **6,** fruiting branchlet, × ⅔. 1–3, from *Paulo* 733; 4, 5, from *Bally* 8965; 6, from *Semsei* 3085.

Twigs puberulous to tomentose; spines 0·4–3·5 cm. long. Leaf-blades coriaceous, elliptic to elliptic-lanceolate, (2·5–)6–11 cm. long, 2·8–5 cm. wide, shortly acuminate and ± long-mucronate, broadly cuneate to rounded or a little cordate at the base, unequal-sided, entire or serrate, glabrous above, glabrous to densely pubescent beneath; petiole 3–6 mm. long. Stipules oblong-lanceolate, 12–20 mm. long, caducous. Cymes of ♂ flowers 5–15 mm. long. Sepals ± 3 mm. long in ♂ flowers, 1–2 mm. long in ♀ flowers, pubescent. Ovary pubescent; styles up to 30 mm. long, persistent, but easily dislodged when dry. Fruits waxy yellow to orange, oblong-ovoid, 12–14 mm. long, 11–12 mm. across. Fig. 4, p. 13.

UGANDA. Karamoja District: Moroto, *Eggeling* 2966!; Ankole District: Bunyaruguru, July 1939, *Purseglove* 884!; Mengo District: Kasa Forest, 5 Oct. 1949, *Dawkins* 430! & 431!

KENYA. Northern Frontier Province: Leroghi [Lorogi] Plateau, 17 June 1959, *Kerfoot* 1070!; Nairobi, 13 May 1953, *Bally* 8965!; N. Kavirondo District: W. Kakamega Forest Reserve, 9 July 1960, *Paulo* 514!

TANGANYIKA. Biharamulo/Ngara District: Rusumo, 23 June 1960, *Tanner* 5006!; Moshi District: Rau Forest, 25 Feb. 1953, *Drummond & Hemsley* 1315!; Lushoto District: Makuyuni, 16 Dec. 1935, *Koritschoner* 1462!

DISTR. **U**1–4; **K**1, 3–6; **T**1–3, 5; Sudan and Congo Republics to West Africa, also Madagascar.

HAB. Understory and margins of lowland and upland rain-forest, also riverine forest; 900–2100 m.

SYN. *C. serrata* Engl. in N.B.G.B. 3: 24 (1900) & V.E. 3(1): 16 (1915); Rendle in F.T.A. 6(2): 14 (1916); F.D.O.-A. 2: 61 (1932). Types: Tanganyika, Usambara Mts., *Holst* 505 (B, syn.) & South Africa, Cape Province, *Beyrich* 119 (B, syn.) & *Bachmann* 432 (B, syn.!) & 433 (B, syn.)

Bosqueia spinosa Engl. in E.J. 40: 548 (1908). Type: Tanganyika, Mwanza District, Ukerewe I., *Uhlig* V73 (B, holo.!)

Chaetacme microcarpa Rendle in F.T.A. 6(2): 13 (1916); F.W.T.A. 1: 423 (1928); F.D.O.-A. 2: 61 (1932); T.S.K.: 85 (1936); I.T.U.: 247 (1940); Hauman in F.C.B. 1: 51 (1948). Type: Sudan Republic, Bahr el Ghazal, *Schweinfurth* 2828 (K, isolecto.!)

VARIATION. In East Africa the species appears to be almost always dioecious, but specimens from South Africa are often monoecious, perhaps a physiological difference related to day-length.

The leaves are occasionally serrate or coarsely spine-toothed, especially on sucker shoots, but this appears to be no more than an individual variation recorded from most parts of the range. There is also considerable variation in the indumentum—one local race of very tomentose plants being recorded from Rau Forest, Moshi District, Tanganyika, e.g. *Drummond & Hemsley* 1315! & *Greenway* 4528! & *H. A. Lewis* 33!

INDEX TO ULMACEAE